About Post Carbon Institute

Post Carbon Institute's mission is to lead the transition to a more resilient, equitable, and sustainable world by providing individuals and communities with the resources needed to understand and respond to the interrelated economic, energy, ecological, and equity crises of the 21st century. **postcarbon.org | resilience.org**

Acknowledgments

Author: Daniel Lerch
Editorial Committee: Asher Miller, Ken White, Richard Heinberg

Thank you to our reviewers, who gave invaluable feedback on various stages of this project; see Appendix for complete list. Thank you also to the funders of this project, including DJR Trust, Lydia B. Stokes Foundation, Threshold Foundation, Proteus Fund, New Visions Foundation, Appleby Foundation, Satterberg Foundation, Alfalfa Fund, and of course, P & L.

See Appendix for image credits.

Six Foundations for Building Community Resilience
 Published November 2015.

post carbon institute

Post Carbon Institute
613 Fourth Street, Suite 208
Santa Rosa, California, 95404

Executive Summary

Efforts to build community resilience often focus on growing the capacity to "bounce back" from disruptions, like those caused by climate change. But climate change is not the only crisis we face, nor is preparing for disruption the only way to build resilience. Truly robust community resilience should do more. It should engage and benefit all community members, and consider all the challenges the community faces—from rising sea levels to a lack of living wage jobs. And it should be grounded in resilience science, which tells us how complex systems—like human communities—can adapt and persist through changing circumstances. *Six Foundations for Building Community Resilience* describes how communities can approach the full scope of the 21st century's challenges equitably and sustainably.

What is the problem we're trying to solve?

Global interconnection is the dominant factor of our modern world. If the aim of community resilience—at minimum—is to safeguard the health and well-being of people in the face of the 21st century's many complex challenges, those challenges need to be understood in a global context. We organize them as a set of four distinct but intertwined "E^4" crises: ecological, energy, economic, and equity. Community resilience building should aim to keep the community from irrevocably changing for the worse as the result of these crises—and ideally change the community for the better.

What is resilience, really?

Resilience is the ability of a system (like a community) to absorb disturbance and still retain basic function and structure. Building resilience means intentionally guiding the system's process of adaptation in an attempt to preserve some qualities and allow others to fade away, all while retaining the essence—or "identity"—of the system. In a human community, identity is essentially determined by what people *value* about where they live. However, what a community of people collectively values is open to interpretation and subject to disagreement. This suggests that people—and the ways they come to rough consensus—are necessarily at the center of community resilience building.

Why communities?

In the United States, state and local governments have significant regulatory and investment power over many of the issues affecting everyday life. This—together with the many ways community members can self-organize and engage in civic life in the U.S.—allows for the kinds of innovations, experimentations, and even failures that are necessary in resilience building, but are not always possible at larger scales. Moreover, because everyone in a community is a stakeholder, it is both ethical and effective for everyone to participate in resilience building and have some responsibility for it: democratic communities have an inherent right to self-determination, and critical community resilience-building processes like social cohesion and system feedback are richest at the local level. Local decision-making doesn't always lead to equitable outcomes, however; one of the weaknesses of decentralization is that parochialism and local prejudice can flourish if unchecked. This suggests two requirements for building community resilience if it is indeed to be organized at the local level:

1. The **responsibility** for resilience building and the power to decide how it is done must rest with community members.

2. The **process** of resilience building must equitably address both the particular situation of the community and the broader challenges facing society.

The Six Foundations

Although many resilience frameworks and tools for building community resilience are now available, no single approach will likely work for all communities and their varied social and economic contexts. Therefore we have identified six foundations that, in our view, are essential—no matter where or how resilience-building efforts are undertaken, or which challenges are of most concern locally. The foundations support *building* community resilience, rather than achieving resilience as a fixed goal, so as to emphasize resilience building as an ongoing process.

The six foundations are:

1. **People.** The power to envision the future of the community and build its resilience resides with community members.

2. **Systems thinking.** Systems thinking is essential for understanding the complex, interrelated crises now unfolding and what they mean for our similarly complex communities.

3. **Adaptability.** A community that adapts to change is resilient. But because communities and the challenges we face are dynamic, adaptation is an ongoing process.

4. **Transformability.** Some challenges are so big that it's not possible for the community to simply adapt; fundamental, transformative changes may be necessary.

5. **Sustainability.** Community resilience is not sustainable if it serves only us, and only now; it needs to work for other communities, future generations, and the ecosystems on which we all depend.

6. **Courage.** As individuals and as a community, we need courage to confront challenging issues and take responsibility for our collective future.

Contents

Preface

Communities across the United States are talking more and more about resilience. They're spurred by recent natural disasters like Hurricanes Katrina and Sandy, weather extremes like the harsh Northeast winter of 2014-15, and long-term drought in the West.

Many people think of a community's resilience as its ability to "bounce back" from disruption, and efforts to build resilience often focus on the impacts of climate change. Climate change is indeed an urgent and existential threat, with untold potential to destroy and disrupt countless lives. But it is not the only crisis we face, nor is preparing for disruption the only way to build resilience.[1] Truly robust community resilience should do more. It should engage and benefit all community members, and consider all the challenges the community faces—from rising sea levels to a lack of living wage jobs. And it should be grounded in resilience science, which tells us how complex systems—like human communities—can adapt and persist through changing circumstances.

Six Foundations for Building Community Resilience describes how communities can approach the full scope of the 21st century's challenges equitably and sustainably. The report draws on some of the most compelling recent thinking about resilience from academia, sustainability advocacy, and grassroots activism, as well as Post Carbon Institute's prior work.[2] It is intended as an accessible resource for local leaders and activists in the United States, and as a contribution to the larger public conversation about resilience in human communities.

The first half of the report presents the challenges communities are facing, the relevant insights of resilience science, and the case for building resilience at the community level. The second half presents the six foundations we feel are necessary for building effective community resilience: **People**, **Systems Thinking**, **Adaptability**, **Transformability**, **Sustainability**, and **Courage**.

Introduction

Everyone agrees that communities should be resilient to disruption. Whether it's a direct hit from a hurricane or the town's main employer shutting down, emergencies happen. That's why we have fire and police departments, EMT services, and an insurance industry.

These days, however, the challenges facing communities have increasingly complex origins. Climate change is making extreme weather events more powerful and less predictable; globalization is tying local needs to faraway economic and political decisions. It seems that in addition to isolated disruptions, community resilience should also reckon with long-term, abstract uncertainties that may emerge and shift without warning.

But wait… communities can't plan for *every* contingency. How do we decide which futures to prepare for, and which predictions to trust? For that matter, *what* exactly in the community should we make more resilient—and how? Should we build higher sea walls to protect the downtown from stronger storm surges, or relocate residents of low-lying neighborhoods and convert those areas to floodplains? Should we give tax breaks to prop up a struggling factory, or let the factory die and try attracting a new local employer with different tax breaks? Who decides, who benefits, and who pays for these decisions that we hope will make the community more resilient?

Wait again: Why focus on potential future disruptions at all when people are facing real and urgent challenges *right now*? In many communities, wages are stagnating and gentrification is pushing long-time residents out of their homes. Whole neighborhoods feel threatened by poverty and crime on one side and institutionalized violence and disenfranchisement on the other. Who wants a "resilient" return to the status quo when the status quo is exactly what needs to change?

Clearly if community resilience is to be a useful concept, we need to take a hard look at what it really means, how it is supposed to work, and what problem it is intended to solve.

What's the problem we're trying to solve?

Virtually every American community is part of—and dependent on—a deeply interconnected and highly complex global civilization of nearly 200 countries, tens of thousands of cities, and over seven billion people. The prices we pay at the grocery store and the gas station, the investments our businesses make, the regulations our governments set, and even the weather we experience every day are potentially influenced by countless events and decisions made around the world—and all to a degree that was barely conceivable just half a century ago.

Although many of the challenges our communities face would exist regardless, this global interconnection is the dominant factor of our modern world and brings us rewards and risks (neither of which are distributed equally) that we cannot ignore. If the aim of community resilience—at minimum—is to safeguard the health and well-being of people in the face of the 21st century's many complex challenges, those challenges need to be understood in a global context.

At Post Carbon Institute we organize those challenges as a set of four distinct but intertwined crises; we call them the "E^4" crises. They influence and multiply each other, and they manifest in myriad ways from the most local to the most global of scales. They are characterized as *crises* because they are pushing us towards decisive changes—tipping points that we may choose to fight, ignore, or take advantage of. The E^4 crises do not encompass all of the challenges facing humanity today, but they frame and highlight those that we feel most immediately threaten modern civilization.

1. **The Ecological Crisis.** Everything we need to survive—to have life, a society, an economy—ultimately depends on the natural world. But every ecosystem has two important limiting factors: its rate of replenishment and its capacity to deal with wastes and stress. The last 200 years of exponential economic growth and population growth have pushed ecosystems around the world near or past these limits, with results like severe topsoil loss, freshwater depletion, biodiversity loss, and climate change. Humanity's "ecological footprint" is now larger than what the planet can sustainably handle, and we are crossing key boundaries beyond which human civilization literally may not be able to continue.[3]

2. **The Energy Crisis.** The era of easy fossil fuels is over, leading the energy industry to resort to extreme measures like tar sands mining, mountaintop removal coal mining, fracking for shale gas and tight oil, and deepwater drilling. But these practices come with significant costs and risks, and in most instances provide far less net energy than the conventional oil, coal, and natural gas that fueled the 20th century.[4] Renewable energy is a real but imperfect alternative, as it would take decades and many trillions of dollars to scale up deployment to all sectors of the economy and retrofit transportation and industrial infrastructure accordingly.[5] Declines in the amount of affordable energy available to society threaten to create major environmental, economic, and social impacts as the 21st century progresses.[6]

3. **The Economic Crisis.** Our local, national, and global economies are currently structured to require constant growth. And yet, with the onset of the Great Recession in 2008, we reached the end of economic growth as we've known it.[7] Despite unprecedented interventions on the part of central banks and governments, economic recovery in the U.S. and Europe has failed to benefit the majority of citizens.[8] The end of the age of cheap and easy energy, the vast mountains of both private and public debt that we have incurred, and the snowballing costs of climate change impacts are all forcing us into an as-yet-undefined post-growth economic system…whether we are ready for it or not.

4. **The Equity Crisis.** Inequity has been a problem throughout recorded human history, and not least in the United States, despite its professed values of liberty and justice for all. While social progress over the last 150 years has in theory brought political enfranchisement and legal protections to almost everyone, in practice the failure to fully extend both economic opportunity and a functional social safety net—together with the failure to fully address institutionalized racism, sexism, and other forms of prejudice—has led to ongoing inequality of economic, social, and political power. The ecological, energy, and economic crises are together exacerbating inequality, which has become increasingly visible in the rapid concentration of wealth among the ultra-rich and in the increasing public anger about police violence against people of color.

These four crises shape the many and complex challenges communities in the United States must wrestle with in the 21st century.[9]

By building community resilience, we are trying to keep the community from irrevocably changing for the worse as the result of these crises—and hopefully change the community for the better. But *how* we go about this is critical to whether our efforts will succeed and last. To understand why, we need to take a close look at the concept of resilience itself.

What is resilience, really?

Resilience is often thought of as the ability to withstand hard times or "bounce back" from a disaster: a town devastated by a tornado is called resilient when its people and its infrastructure are able to quickly return to how things were before.

People working on community sustainability issues have developed a more nuanced view of resilience over the last fifteen years. A commonly used approach—and the one used in this report—comes from the field of ecology, where resilience is understood as the ability to absorb disturbance and still retain basic function and structure, or "identity."[10] In other words, a resilient system can adapt to changes without losing the essential qualities that define what it is and what it does. For example, a maple-beech forest ecosystem might experience wildfire, drought, or infestation; but if it is sufficiently resilient it will recuperate from individual incidents and adapt to longer-term changes, all while keeping essentially the same species, patterns, and other qualities that define its identity of "maple-beech forest ecosystem."

In resilience science, a community and the ecosystem it makes use of are together considered a unified socio-ecological system. The system's adaptability is a function of general characteristics like diversity, innovation, and feedback, as well as its ability to cope with vulnerabilities specific to its situation and to make deeper transformations if needed.[11] Importantly, the system is understood to be a "complex adaptive system" that is not static but is constantly adapting to change—change that is often unpredictable.[12]

When we intervene in a system with the aim of building its resilience, we are intentionally guiding the process of adaptation in an attempt to preserve some qualities and to allow others to fade away—all while retaining the essential nature, or "identity," of the system. Thus, resilience building necessarily starts with decisions about *what we value*. Of course, what a community can be said to "value" is open to interpretation and may not be agreed upon by everyone. It may even reflect ignorance and prejudice; few today would agree with racist and sexist values dominant in many U.S. communities even 50 years ago. As we'll see later, these core issues of equity and values make a people-centered approach to community resilience especially important.

Resilience science has mostly focused on rural communities and the natural resources they depend on (see sidebar, "Resilience thinking in action")—but new efforts are exploring how it can be applied to non-rural communities and their relationships not only with ecological systems but with economic and social systems as well.[13] We might ask, for example, how a city can address complex challenges like a globalizing economy, more frequent extreme weather, rising healthcare costs, and uncertainty about the future mix of energy resources.

Applying resilience thinking to a modern city isn't fundamentally different from applying it to a small rural community: we are simply considering a broader scope of systems because it is within that community's power to do so. A mid-sized American city has billions of dollars in infrastructure and social spending to work with over multiple years, not to mention hundreds of thousands of people who can act towards various goals through their economic, civic, and social activities. (Of course, the challenge of facilitating decision-making among the larger community's competing interest groups will be more complex.)

When applied to communities, resilience is sometimes spoken of as the "next generation" of sustainability; and indeed, our definition of community resilience (see page 10) deliberately incorporates sustainability's nested triad of environment, society, and economy. But the two concepts—*resilience* and *sustainability*—may also be understood as different frameworks for achieving the same goal: organizing how we interact with the world around us and with each other in ways that can continue indefinitely. Sustainability thinking has made important contributions to how we value and steward the resources our communities depend on, although its aspirations have proven difficult to put into meaningful practice at large scale.[14] Resilience thinking offers a complement to sustainability thinking in that it is explicitly focused on the challenges of humans coexisting with ecological systems—it was developed for practical use in the messy, unpredictable real world.[15] As Charles Redman of Arizona State University has put it, "sustainability prioritizes outcomes; resilience prioritizes process."[16] (We will discuss this relationship later in the report.)

Resilience can be a powerful concept for communities…but why bother building resilience at the community level at all when the E^4 crises are ultimately national and global in scale? We'll see why in the next section.

Sidebar: Resilience thinking in action

Imagine a farming community faced with declining crop yields caused by decades of poor soil management, misplaced government priorities, and long-term declining rainfall. Most of the residents recognize that things need to change, but they don't want to lose the things about their community they most cherish: their deep connection to the land, the generations-long relationships between neighbors and families, and shared ethics like looking out for each other and maintaining some capacity of self-reliance.

If this community invited in a resilience scientist, she might lead them through a process that looks like this:

- They would start by describing the entire socio-ecological system of the human community together with the ecological system it makes use of and depends on. The stakeholders both within and outside the community—farmers, residents, government managers, environmental advocates—would gather and talk about what defines that combined socio-ecological system and how it works, finding agreement on important factors and dynamics.

- They would look at attributes of the system that contribute to its capacity to cope with change. They'd pay close attention to areas of the system that may be particularly vulnerable, where too much change could have unwanted or even irreversible effects. Perhaps yields of the community's historically dominant crop are declining because of topsoil loss and declining rainfall, but farmers feel trapped into growing it because of past investments and government incentives, and thus are increasingly vulnerable to price and weather fluctuations.

- They would develop plans to both cultivate overall resilience with specific interventions and flexibly manage the system over the long term. Perhaps the farmers, business owners, and government officials form a local stewardship council to monitor, improve, and coordinate farming practices, and agree on a five-year plan to help farmers transition some land to different uses, support businesses in providing the new services needed, and train young people in a greater diversity of skills.

This simplified example includes some critical aspects of resilience thinking. It deals with complexity: multiple factors both inside and outside the system are interacting in ways that aren't always predictable. It deals with identity: the community values certain essential aspects of what it is and how it works, and wants to retain them in the face of change. And it deals with local stakeholders and power: the path to an effective and equitable solution depends on people in the community not just having a voice but actively participating in decisions and management.

Why communities?

When we speak of a "community," we mean something far more than just the physical infrastructure of a human settlement. A community is also the people inhabiting a particular place, defined by their interpersonal relationships, cultural patterns, economic and governance structures, and shared memories and aspirations.

In this report "community" is left loosely defined, envisioned as a place-based group of people who have some meaningful capacity to influence their basic common needs given their particular social and political context. In urban areas it might be a city of a few million with all its competing interest groups, or a close-knit neighborhood of just a few thousand. In rural areas it might be a village of a few hundred, or a 5,000-square-mile county of dispersed towns. Community resilience building can start with whatever scale and set of people the initiators deem appropriate in a given situation—although through discussing needs, aspirations, and capacity (with attention to the six foundations presented in this report) it should quickly become apparent if the scale should be expanded or contracted.

The argument for building community resilience—and specifically for doing the work at the community level—is twofold. First, in the United States, community-level resilience building makes practical sense because of how our political system is structured. By design, new ideas typically come to fruition at the federal level slowly, thanks in part to the roles and constraints set by the Constitution and the procedural hurdles of Congress. In contrast, local and state governments often have great flexibility in organizing how public decisions are made, as well as significant regulatory and investment power over the issues that most affect everyday life: social services like health care and police; public goods like utilities; civic institutions like schools and courts; land use and transportation planning; and so on.

Indeed, our cities and states are traditionally the country's laboratories for social and economic innovation.[17] One community's experiment can inspire thousands of other experiments, providing valuable insights and best practices, and ultimately building support for larger-scale changes. During the previous decade, while national and international climate efforts languished, cities across the country followed early leaders like San Francisco and Seattle and started their own climate initiatives. Using the terminology of resilience science, we might say that cities and states are providers of diversity, openness, and modularity for the resilience of the higher-level national system.

This model of local innovation works as well as it does because it is at the community level where we (as individuals, businesses, organizations) most directly interact with the people and institutions that make up our society. It is where we are most affected by the decisions society makes: what jobs are available to us, what infrastructure is available for our use, what policies exist that limit or empower us. And critically, it is where the majority of us who do not wield major political or economic power can most directly affect society: as voters, neighbors, entrepreneurs, consumers, activists, and elected officials.[18]

From that observation arises the second part of the argument for building resilience at the community level: it is both ethical and practical for community members to be at the heart of community resilience building work. (This may seem self-evident but it isn't necessarily so; a central government attempting to direct the resilience-building efforts of thousands of communities remotely, relying on uniform indicators, outside managers, and centralized resources.) Using the terminology of sociology, we might say that everyone in a community is a stakeholder—and those stakeholders need the opportunity not only to participate in resilience building but to actually have some responsibility for it.

Decades of research underline how important it is for local stakeholders to have real power in decisions that affect them.[19] Some of the central concepts of resilience science tell us why this particularly applies to urban communities. For example: *identity* (as discussed earlier) is the touchstone of a system, and in a democratic society the members of a community have an inherent right to self-determination; hence the identity of the community emerges from its members. *Social capital*—people's relationships—is what gets things done in human systems, and is richest at the local level. Local connections and presence also create more and tighter opportunities for system *feedback*, which is essential for adaptation and innovation.

For us as social animals, identity is tied to community: our relationships to other people and to a place; our sense of shared experience, history and culture; the smells and sounds and even the soil that we associate with "home." How else can community members recognize themselves as stakeholders if not by seeing themselves as part of a larger place-based whole?

The ability to put local stakeholders at the heart of resilience-building efforts, plus the practical advantages of community-level government in the United States, make community resilience building an effective way to respond to the E^4 crises. But local decision making doesn't always lead to equitable outcomes; indeed, one of the weaknesses of decentralization is that parochialism and local prejudice can flourish if unchecked. This suggests two requirements for building community resilience:

1. The **responsibility** for resilience building and the power to decide how it is done must ultimately rest with community members.

2. The **process** of resilience building must equitably address both the particular situation of the community and the broader challenges facing society.

These requirements—in dynamic tension with each other, because together they task community members with acting beyond their own self-interest—are the starting point for the six foundations of building community resilience, described in the remainder of the report.

+ + +

> How do you know community resilience when you see it? I think you look for the capacity for people to not have to go through extremes... being knowledgeable and having capacity to do something, to change your circumstances.
>
> —Doria Robinson, Urban Tilth[20]

> We all need a sense of community. And we all need to believe that we have agency—a sense that we can make choices that will affect our lives.
>
> —Stuart Comstock-Gay, Vermont Community Foundation[21]

> We all come from cultures that built deep resilience because we were able to exist in the place we lived for a long time. We call that cultural diversity. This evolved knowledge of place. In a way, we look back to our indigenous ancestral wisdom to see models of how people who got to live in one place for hundreds of years really knew how to take care of that place, that home. Resiliency is there. It's in all of us.
>
> —Ellen Choy, Movement Generation[22]

The Six Foundations

Although many resilience frameworks and tools are now available, no single approach will likely work for all communities and their varied social and economic contexts. Therefore we have identified six foundations that, in our view, are essential for community resilience—no matter where or how resilience-building efforts are undertaken, or which challenges are of most concern locally. We define community resilience as *the ability of a community to maintain and evolve its identity in the face of both short-term and long-term changes while cultivating environmental, social, and economic sustainability.*

The six foundations are intended to provide a theoretical resource for advocates, activists, and local leaders working to make their communities more resilient—in many cases with practical efforts that are already underway but that need to be replicated, strengthened, and supported. They are derived from existing resilience frameworks and principles, interviews with advocates and activists working on resilience-related issues across the country, and Post Carbon Institute's own past work (see Appendix).

The foundations support *building* community resilience, rather than resilience as a fixed goal, so as to emphasize resilience building as an ongoing process. While an individual initiative may not be able to build upon all of these foundations, in our view a community's overall resilience-building effort must include all of them in order to be effective.

The six foundations of building community resilience are:

1. **People**.
2. **Systems thinking**.
3. **Adaptability**.
4. **Transformability**.
5. **Sustainability**.
6. **Courage**.

Foundation #1: People

The power to envision the future of the community and build its resilience resides with community members.

> We can try to outsource our problems to a new generation of green engineers, designers, and architects, but we will only see broad, lasting changes when the people inhabiting these communities create a vision for the future and lead the process for change.
>
> —Phil Myrick, Project for Public Spaces[23]

WHAT IT MEANS

Communities are products of human relationships.[24] What the community is now and what it will be in the future both result from decisions made by people interacting, negotiating, and working together. Trust and deep relationships are crucial to holding communities together year after year and making resilience durable—but they can be challenging to build, especially in diverse communities.[25]

Resilience building cannot turn a blind eye to the political and economic processes that determine what gets done, how it gets done, who decides, and who benefits. People of all interests and means must be able to participate in and benefit from resilience building; indeed, if they are to build true resilience, communities must *embrace* dissent and diversity.

The goals of community resilience-building efforts are best set by and focused on the needs of the people who make up the community—not just the needs of the most politically engaged or powerful individuals, businesses, and external stakeholders. Also, community members must collectively have power and responsibility for cultivating the resilience of their community as active participants and leaders—rather than only the local government or business leaders holding power and responsibility.

WHY IT'S IMPORTANT

Identity

Resilience is the ability of a system to deal with disruption and change while keeping its basic functions and structure—its "identity." In a democratic society we might say that the identity of a community arises from its members and represents a shared sense of what the community's core qualities are. And because we humans have aspirations and free will, we might also say that identity includes a shared vision of what the community should be like in the future. We can try to describe a community's identity by asking people: What are the values of this community? What defines this community, and why? What do we not want to lose? What do we need to change? These kinds of questions can only really be answered by community members.

Identity is the touchstone of a community's resilience. But as an expression of values, it also shapes perceptions of what is important and what is worth doing. This suggests that for the work of resilience building, the *way in which* identity is characterized is quite important.[26] A few considerations:

- Systems are defined by their larger context—and human communities exist within larger social, economic, and ecological systems. So, the voices of outside stakeholders and experts are important to prevent parochialism and include specialist knowledge.[27]

- Systems are also defined by their components—and human communities are aggregates of smaller social groups with varying levels of influence and power. So, the voices of traditionally disempowered or dissenting groups are not only ethically important to include, they can also help prevent discrimination and stagnation (although this responsibility is shared by all).

- In human communities, identity is dynamic.[28] It is a function of people existing in a community together, changing as they and the society and environment around them change. University of Colorado professor Bruce Goldstein notes that "identity and community are collaborative achievements, not just entities already out there waiting to be found and dusted off."[29] Resilience-building efforts should constantly revisit and refine their understanding of what the community's identity is.

In practice, envisioning a shared community identity will be messy, multi-faceted, and constantly open to question.[30] Opening potentially challenging discussions is essential for uncovering not only inequities and vulnerabilities, but also opportunities and resources.

Effectiveness

Resilience building is most effective when stakeholders are engaged and invested—and in communities, the primary stakeholders are the people who live there. The people living within a community are the key to the crucial resource of social capital—essentially, the local relationships that make things happen.[31] They are often the most knowledgeable about the community's opportunities and challenges, and best-suited to act on them through existing economic, political, and social relationships.[32]

When community members have ownership of and responsibility for resilience building it creates a sense of agency and support for the work—as well as of fairness and shared effort in what emerges. (Indeed, this is partly why resilience building can't just be a government project.[33]) It helps with the longer, broader process of social cohesion—the formation of bonds that make us willing and able to cooperate, collaborate, and take care of each other. Social cohesion is essential for helping us get through acute crises like natural disasters,[34] and makes a community feel enriching and nurturing over the long-term.

Social capital accumulates and evolves over time, allowing the community to continually build up its knowledge, skills and place-based wisdom—things that so many communities have lost over the last century.[35] It's more than a renewable resource; the more we use it, the more it grows, and the more it contributes to community resilience.

Foundation #2: Systems thinking

Systems thinking is essential for understanding the complex, interrelated crises now unfolding and what they mean for our similarly complex communities.

I have seen repeatedly that a too-narrow understanding of the issue—from only limited vantage points or within only one sector for example—leads to poorly framed interventions. Thinking in systems goes beyond any one segment or sector and pushes groups to include those "unlikely bedfellows" that can help find the leverage points for change.

—Michelle Colussi, Canadian Centre for Community Renewal[36]

What are you trying to do, and what are the consequences? To me that's systems thinking. It's thinking about how one action here affects the whole. It's taking responsibility for taking actions.

—Doria Robinson, Urban Tilth[37]

WHAT IT MEANS

Our communities are thoroughly integrated sub-systems of a single global socio-ecological system. They're connected to or influenced by external factors like regional water supplies, national energy policy, and global climate change. Our communities are also complex systems in their own right, with innumerable components constantly changing and interacting with each other, the larger whole, and outside systems. Local economic activity, relationships among different social groups, local cultural patterns... they all influence the community from the inside out.

The challenges we face are complex, so we can't approach them as if they were linear problems. Systems thinking helps us understand the complex E^4 crises, as well as how our complex societies and communities work. It is also the basis of resilience science.

WHY IT'S IMPORTANT

Making sense of complexity

Systems thinking—simultaneously seeing the parts, the whole, and the relationships within a system[38]—helps us make sense of complexity. Complexity is different from being complicated. Resilience thinkers Brian Walker and David Salt describe it this way:

> The mechanism that drives an old-style clock is a set of tiny, intricate cogs and springs, often consisting of many pieces. This is a complicated machine... However, the individual pieces are not independent of one another; rather, the movement of one depends on another in an unvarying way... [In contrast,] although a farm might produce just one item (e.g., wheat), the farm is far from simple. The farmer, the farming practices, the crop, the soil it grows on, and the market are all interacting and changing over time. This is a complex adaptive system.[39]

Engineering helps us understand the clock but it will only get us so far with the farm. Weather, market prices, soil nutrition, government policy and countless other factors are all in flux and often unpredictable. Systems thinking gives us concepts that help us model the dynamics and relationships that exist. We can start to think of the farm in terms of "stocks" (resources like the wheat in the storehouse; the nutrients in the soil), "flows" (sales of the wheat; depletion of the soil's nutrients), "feedback loops" (higher demand for grain spurs the farmer to plant more wheat; more cultivation means the farmer needs to replace more lost soil nutrients), and so on.

An essential part of systems thinking is setting a boundary: deciding the limits of what we'll consider in detail. By setting a boundary, we are not pretending that everything outside the boundary doesn't exist—rather, we are choosing one of many possible perspectives, and accepting that we can't know everything we might want to know. Indeed, recognizing that there is more than one way to see things is at the heart of systems thinking. This is especially important when we are talking about human communities, where there is rarely a lack of diverse views and interests.

If we'll never have complete information, it follows that there will always be blind spots. During the run-up to the Iraq War, U.S. Defense Secretary Donald Rumsfeld famously described this as the problem of "known unknowns" and "unknown unknowns."[40] This suggests that an open-ended, adaptable response to a problem may be preferable to a static solution. As we'll see with the next foundation, Adaptability, resilience science gives us tools for anticipating and dealing with uncertainty.

Making the E^4 crises relevant

Modern industrial society operates today at a global scale, and every community is deeply dependent on resources and processes far beyond its own region.[41] International trade and relations are of course nothing new, but over the last half century we have created extraordinarily complex interconnections between economic, social, and environmental systems around the world. Building community resilience in the face of the E^4 crises means we need to think about the myriad challenges (of which only some are predictable) that we'll face in the foreseeable future.

Consider, for example, the complex relationship that U.S. communities have with fossil fuels and climate change. Communities currently rely on fossil fuels to provide essential energy services—fuel for vehicles, agricultural inputs, heat for buildings and industrial processes, electricity for communications, and more. However, our communities' dependence on fossil fuels is a major driver of climate change, both directly (burning fossil fuels for transportation, electricity, and heat) and indirectly (consuming food and goods manufactured and transported with fossil fuels). Climate change is, in turn, affecting our communities—also directly and indirectly. The direct impacts are obvious and much-discussed: storm surges that damage buildings and infrastructure; droughts that reduce local water supply; extreme heat and cold that endanger vulnerable populations; and more. The indirect impacts are less obvious: drought in one part of the world (or even one part of the country) might hurt agricultural production and cause food prices to rise elsewhere; climate-driven economic and social unrest in an oil-producing country might disrupt exports, impacting the price or supply of gasoline.

Understanding the E^4 crises can help guide actions at the community level. For example, if we assume that the market will automatically supply affordable energy as long as there is demand, there is no point in worrying about the trend of diminishing cheap-to-produce oil resources. On the other hand, when we understand the basic mechanisms of our energy crisis—i.e., that our economy and infrastructure remain extremely dependent on oil, and alternative energy sources are all limited in their capacity to substitute for it[42]—we get a better sense of what to expect in the future and what it might mean for our community.

Systems thinking makes the E^4 crises relevant to communities in one other way: It helps us see that actions even at the relatively small community level play a role in what is happening at the national and global levels. They are all parts of the same system. Building community resilience contributes to the resilience of our global socio-ecological system.[43]

Foundation #3: Adaptability

A community that adapts to change is resilient. But because communities and the challenges we face are dynamic, adaptation is an ongoing process.

> In a time of drastic change it is the learners who inherit the future. The learned usually find themselves equipped to live in a world that no longer exists.
>
> —Eric Hoffer, *Reflections on the Human Condition*[A4]

WHAT IT MEANS

When complex systems are resilient in the face of disruption it is because they have the capacity to adapt to changing circumstances, thanks to system characteristics like diversity, modularity and openness. In human systems, resilience-building efforts aim (in part) to cultivate such characteristics—but if those efforts themselves don't adapt to changing circumstances, they may unwittingly cultivate the resilience of things that *aren't* desired. (Poverty, drought, and authoritarian governments can all be resilient in their own ways.)

WHY IT'S IMPORTANT

The qualities of resilience

There are many different ways to think about how resilience is built and how adaptability is supported. In their influential book *Resilience Practice*, Brian Walker and David Salt list "attributes" like *diversity*, *modularity*, *openness*, and *reserves*.[45] The Stockholm Resilience Institute identifies "principles" like *manage connectivity* and *broaden participation*.[46] The Rockefeller Foundation lists "qualities" like *robust*, *redundant*, *flexible*, and *inclusive*.[47] While some of these terms and approaches differ, they essentially point to the same ideas. For communities, what matters is that resilience is understood as a quality to continually cultivate by taking on the right patterns, not a goal to be achieved by ticking off a list of characteristics. Andrew Zolli (author of *Resilience: Why Things Bounce Back*) evokes this approach with his "verbs of resilience"—four things that are happening all the time in a resilient community:

1. **Building** regenerative capacity.
2. **Sensing** emerging risks.
3. **Responding** to disruption.
4. **Learning** and transforming.[48]

Initiatives, activists, and politicians come and go, but if resilience building is ingrained in the community culture, it can evolve as the community evolves.

Learning

Adaptability is both about responding to change (both external and internal) and learning from the experience. Learning happens through feedback loops. In a model system, feedback loops send information from one part of the system to another so that it can self-regulate; resilience is built by having tight feedback loops. A community lacking in resilience is probably suffering from poor or incomplete feedback loops: perhaps community members don't know what business and government leaders are doing, or certain groups of people don't have a voice in the community. Effective resilience building aims to identify what types of feedback (and from where and to where) are important, including those that are being overlooked or ignored.

The problems of complexity and efficiency

The adaptability of a system is influenced by many things, and often not in obvious ways. For example, too much complexity in a system can be a symptom of low resilience: it can reduce flexibility and create resistance to change. In the northeastern blackout of 2003, a few minor problems in Ohio suddenly overwhelmed the electricity distribution system's ability to cope, causing a massive power failure affecting 55 million people. The physical system was no longer able to adapt because it had too much complexity.

One way to potentially reduce excess complexity is to improve efficiency—but this can also have unintended consequences. For example, the post-World War II push to move poor families into oversized, anonymous public housing projects was deemed an efficient way to provide housing cheaply. But it also cut the rich social ties and emotional roots people had in their old neighborhoods, making it easier for crime to flourish and destroying the social capital that might have been tapped to address community challenges.[49] These "planned" social systems were less able to adapt because they had *too little* complexity.

Too much resilience

Communities, their subsystems, and the systems they are part of are all constantly changing, and in ways that are often unpredictable. A system that cannot cope with change will ultimately cease to exist. The collapse of the Soviet Union may be the most dramatic example in living memory of a human system whose failure to adapt to both external and internal changes proved fatal.

In contrast, the U.S. political and economic system has been quite resilient—largely because of system characteristics that build resilience, like diversity (competition is encouraged), innovation (financial and social incentives exist for profitable ideas), and reserves (when markets fail, governments have stepped in with bailouts). Resilience can become a problem, however, when the decisions that cultivate resilience-building qualities themselves fail to adapt. The severe market failure of 2008 was essentially brought on by the U.S. system's overdependence on debt and cheap oil (which is a complex function of public sector policies and private sector investments).[50] Economic collapse was avoided, but at the cost of actions that ultimately *reinforced* dependence on debt and oil—that is, the system achieved short-term stability but increased its long-term vulnerability. Unless the system can "learn" and truly adapt to the changed reality (i.e., stagnant real economic growth and the end of cheap and easy fossil fuels), it may not get through the next crisis without deep—and likely undesirable—transformation.[51]

Foundation #4: Transformability

Some challenges are so big that it's not possible for the community to simply adapt; fundamental, transformative changes may be necessary.

The way you maintain the resilience of a system is by allowing it to probe its boundaries.

—Brian Walker, resilience scientist[52]

If we want things to stay as they are, things will have to change.

—character in *The Leopard* by Giuseppe di Lampedusa[53]

WHAT IT MEANS

Communities generally adapt as the world around them changes. But if adaptation happens too slowly or is constrained, challenges can outpace the ability to cope and eventually threaten overall resilience. When automobile manufacturing started moving out of the Midwest, for example, many communities were so dependent on the industry that mere adaptation wasn't an option: they needed to radically rethink their economic basis (and the social and governance implications of radical change) if they hoped to maintain any ability to chart their futures. In other words, these communities needed to change some part of their identity (while retaining their most valued qualities) and transform to a new state that could be resilient under the new circumstances.

Resilience building usually tries to maintain the basic function and structure of a system in the face of disruption. Transformational efforts are *purposefully* disruptive to the system, changing some of its functions and structures so that it can build resilience in ways more suited to the new reality.

WHY IT'S IMPORTANT

It is hard to get new results from old patterns. Past investments in now-outmoded infrastructure aren't easily abandoned; entrenched leaders rely on existing relationships and hold on to outdated assumptions and prejudices; bureaucracies ossify in decades-old procedures that everybody hates but nobody seems to be able to change.

A system's ability to potentially remake itself—to transform—is a key component of its overall resilience (the other components are its general adaptive capacity and its ability to cope with vulnerabilities specific to its situation).[54] In some situations it may be necessary for the entire system to transform. In the 1990s, the Austrian community of Güssing transformed itself from a poor agricultural town into a minor industrial center by completely remaking its relationship with energy, going from importing all of its (mostly fossil fuel) energy to becoming a net renewable energy producer.[55]

In other situations, it may just be a single but essential part of the system that must transform in order to achieve greater system resilience. Imagine a community police department with an entrenched culture that disproportionately arrests and harms young black men. This essential sub-system of the community—the law enforcement function—is undermining overall resilience by violently disrupting lives and households, feeding resentment towards local authorities, and raising the chances of social unrest. The police department needs a different culture, different internal policies, and possibly different leaders; it needs to transform into something significantly different from what it currently is.

Community resilience-building efforts can be transformational by tackling those aspects of the community that need fundamental change, and sowing the seeds of transformation generally for when change is needed in the future. In resilience science, transformability depends on three attributes:[56]

1. **Getting to acceptance.** Transformation is intentional disruption, so it will not be successful unless the people involved and affected recognize the need for it. Information, transparency, dialogue, and inclusive processes are all important.

2. **Having options for transformational change.** New ideas for dealing with new situations will only be available if there is room for them to be developed and tested. Resilience-building efforts might aim to allow and create space—regulatory, economic, social, and even physical space—for experimentation and novelty within governments, businesses, and neighborhoods, as well as seeking out innovations from the margins (which is where transformational change often starts).

3. **Having capacity for transformational change.** As Brian Walker and David Salt describe it, "transformative change needs support from higher scales and also depends on having high levels of all types of capital—natural, human, built, financial, and social."[57] Support from "higher scales" could mean that state policymakers have good working political relationships with local elected officials; or that there is a solid regional network of charging stations in place to support the city's new electric vehicle program. Of the "high levels" of capital needed, the potential of social capital is particularly compelling; consider, for example, the deep social and cultural relationships that were integral to success of the 1960s Civil Rights movement.[58]

Foundation #5: Sustainability

Community resilience is not sustainable if it serves only us, and only now; it needs to work for other communities, future generations, and the ecosystems on which we all depend.

> For those who embrace sustainability in the fullest sense—as an environmental, social, economic, and political ideal—we're at a crossroads in our civilization. There are two paths to take: continue with business as usual, ignore the science of climate change, and pretend that our economic system isn't on life support—or, remake and redefine our society along the lines of sustainability.
>
> —Jeremy Caradonna, *Sustainability: A History*[59]

WHAT IT MEANS

As discussed earlier in this report (see "What is resilience, really?", page 4), sustainability and resilience are distinct concepts that complement each other. Resilience helps us understand the nuts and bolts of how socio-ecological systems work and how they might adapt (or fail to adapt) to changes over time. Sustainability helps us understand in a more general sense our extremely complex relationship with the natural world, and the consequences of getting that relationship wrong. Where resilience is process-oriented and, in ways, value-neutral, sustainability forces us to confront deep questions and uncomfortable potential futures.

Sustainability is a guiding light for resilience building, where there can be a danger of getting overwhelmed by endless system factors and dynamics. Its tools help us make sense of the torrent of information that systems thinking requires us to explore. The perspective we get from it informs the long-term goals of resilience building. But we also need to be careful in our pursuit of sustainability that we don't mistake what we want for what's actually possible.

WHY IT'S IMPORTANT

Tools

Sustainability starts with the obvious but still often ignored observations that humanity's actions are ultimately limited by the carrying capacity of our finite planetary biosphere, and that we are already running afoul of this limit. In general, it is concerned with exploring how our actions impact the biosphere, how the biosphere in turn impacts us, and how our actions need to change over the long term. Community resilience-building efforts will find useful guidance for grappling with the E^4 crises in certain observations and analytical tools that have been developed in sustainability thinking:

- **Limits to growth.** As Post Carbon Institute's Richard Heinberg notes, "in 1972 the now-classic book *Limits to Growth* explored the consequences for Earth's ecosystems of exponential growth in population, industrialization, pollution, food production, and resource depletion… The underlying premise of the book is irrefutable: At some point in time, humanity's ever-increasing resource consumption will meet the very real limits of a planet with finite natural resources."[60] The related "ecological footprint" concept shows us how humanity is using the Earth's resources faster than it can regenerate them, and challenges us to think about whether everyone can and will get a fair share.[61] Community resilience-building efforts may ask: Are we assuming that economic growth will continue? What does our future look like if the natural resources we depend on become scarcer or more expensive?

- **Capital and services.** Environmental and human resources are often thought of as forms of *capital*—namely, natural capital and social capital—when considering the services and benefits we receive from them: natural capital, perhaps in the form of a forest, can provide *services* like cleaning air and filtering water; social capital includes the relationships found within a community, and is the basis for organized action. Sustainability thinking can help us think about how these and other resources might be valued against each other—and if it is even possible (or ethical) to do so. This has practical implications for communities. For example, if we cut down a nearby forest so that our expanding community has more room for homes and jobs, and we offset the loss by building parks elsewhere, is that a defensible trade-off?[62] If gentrification pushes established long-time residents out of a neighborhood but spurs overall community economic growth, is *that* a defensible trade-off?

- **Safe operating space for humanity.** In 2009, Johan Rockström and colleagues proposed a model of nine planetary boundaries within which humanity must remain to avoid catastrophic environmental change.[63] They include limits on climate change, interference with the nitrogen and phosphorous cycles, biodiversity loss, and ocean acidification. Community resilience-building efforts may ask: Are we contributing to humanity pushing past these boundaries? Are we prepared for catastrophic environmental change? What can we do to reduce our impact—and prepare for the unavoidable changes—locally?

- **Seven generations.** The essential aspiration of sustainability is for human civilization to persist on this planet indefinitely.[64] This suggests two requirements for community resilience-building efforts that do not necessarily emerge from resilience thinking on its own: they must benefit both present and future generations, and future generations must be able to continue them.

A non-negotiable yardstick

Of course, sustainability is far more than a suite of useful tools and a theoretical goal to which we should aspire for the sake of future generations: It presents us with a non-negotiable yardstick against which all our actions, goals, and plans must be measured. Quite simply, these are either sustainable or unsustainable. But rather than face the reality that many of our individual and societal activities—and even our well-intentioned environmental strategies—are incompatible with true sustainability, we've re-appropriated the term to refer to practices that are merely more environmentally sound than others.[65]

How can sustainability, as a way of thinking about the world, remain meaningful if it doesn't seem to be leading us where we urgently need to go?[66] The problem is not the concept of sustainability per se, but rather that we've collectively lacked the courage to engage with it as honestly as needed. We too easily use sustainability to think critically about the present but only optimistically about the future. In the 1990s, when sustainability was first becoming a household word, it evoked shocking images of disappearing rainforests and stranded polar bears—but inevitably with a hint that tragedy could be reversed if only we each did our small part.[67] Two decades later, with the rainforests still burning and the polar bears still starving, it's clear that a more pragmatic and sober approach is overdue.

Such an approach to sustainability recognizes that if we don't find strategies to keep the human project operating within the limits of the biosphere, that project will ultimately fail. It challenges us to confront a damaged future and, even more importantly, to learn from our mistakes so that we stop making things even *worse*. Pragmatic, sober sustainability lends urgency and depth to resilience-building efforts at the community level: We each need to do our part indeed—and it can't be small. There's too much at stake.

Foundation #6: Courage

As individuals and as a community, we need courage to confront challenging issues and take responsibility for our collective future.

> More and more I see people who just know the status quo isn't working—they don't have courage, they just know they need some different answers. Accepting the answers may require courage but if they are engaged in co-creating them, there is ownership and commitment.
>
> —Michelle Colussi, Canadian Centre for Community Renewal[68]

> Hope is...an ability to work for something because it is good, not just because it stands a chance to succeed... It is not the conviction that something will turn out well, but the certainty that something makes sense, regardless of how it turns out.
>
> —Vaclav Havel, *Disturbing the Peace*[69]

WHAT IT MEANS

Community resilience building is not an engineering problem solvable just by knowledge and skill. It is a social undertaking, involving thousands or even millions of people and their most meaningful relationships, hopes, and fears. It confronts us with the worrying threats of the E^4 crises and compels us to engage with people with whom we may disagree—perhaps quite strongly.

We need motivation and emotional strength to take on such personally challenging work. Individuals need courage to speak out about their views and needs, and make themselves personally vulnerable. Communities, too, need courage to create space for difficult conversations, make far-reaching investments and policy changes, and risk sharing political and economic power.

Courage is the ability to do something you know is difficult, and building community resilience in the face of the E^4 crises can be difficult indeed. Resilience-building efforts need to cultivate courage in both individuals and the community as a whole to confront challenging issues and take responsibility for their collective future.

WHY IT'S IMPORTANT

Facing problems head-on

Resilience building makes us grapple with complex problems that don't have easy or obvious answers. It can be overwhelming to try to make sense of the global E^4 crises, not to mention local challenges. Moreover, these are challenges that literally hit close to home. From the daily injustices of the equity crisis to the existential threat of climate change, the E^4 crises threaten our physical, economic, and emotional well-being, as well as some of the things we most hold dear: home, family, friends. These are big, long-lasting that problems will affect our children and grandchildren—as will the actions we take in response to them.

Collaboration isn't easy

It's hard enough to work on these issues as individuals and households; it's harder still to work on them as a community, with people who may see things differently. Take, for example, the challenge of finding basic agreement about the "identity" of the community (see page 12). Should the community aim for growth or stability? Should it preserve the dominant culture, or be open to new people and new ideas? There will inevitably be disagreement and even struggle over such questions, because social change is always negotiated and contested.

Even finding agreement on which problems are most urgent can be contentious. Urban planner Saharnaz Mirzazad recalls participating in a public meeting about community resilience in Oakland, California in 2015: "Gentrification, climate change, and fair wages were all part of the discussion. However, community representatives were more concerned about gentrification than climate risks because that was an immediate threat forcing them out of the community."[70]

Talking seriously about the community's future also means talking about the community's past: How did its current trajectory come to be? This can lead to uncomfortable but important conversations about present and past injustices, and how power is wielded in the community. Although they can be awkward, such conversations open the door to deliberation about how power can be more equitably shared in the community. In fact, if community resilience-building efforts aren't challenging, they're probably not going deep enough.

Sticking with the work

We humans form communities in part because we want stability and predictability. We've evolved systems over millennia to provide us with food and water, enable us to move long distances, and interact with each other without constantly fearing for our safety. Those systems—built infrastructure, social institutions, cultural patterns—are understandably resistant to change. It takes courage to imagine and then do things differently than they've been done before, whether it is adapting current practices or transforming them more fundamentally.

Courage also supports us through the practical challenges of collaboration and public process; logistical obstacles pop up, volunteers disappear, funding runs out, or we simply don't get what we want. It takes courage to collaborate with our neighbors—even on seemingly inconsequential matters. Charla Chamberlain, co-founder of The City Repair Project, tells this story about a neighborhood mural project in Portland, Oregon:

> One of the neighbors at the meeting was an artist, and adamant that the colors of the street painting be a certain way. The discussion became strained, and had not been resolved by when she had to leave. A few days after the meeting, one of her neighbors came to her door. The woman timidly showed her a few sets of colors they had chosen after she'd left, and said she and the group had wanted to be sure the artist was OK with what they'd decided. As the artist related this story to me her eyes welled up with tears, and she told me she realized in that moment that her relationship with her neighbor was far more important than whatever color was chosen.[71]

Whether it is organizing a neighborhood street mural, campaigning for energy efficiency, or fighting institutionalized racism, getting involved with your community and making yourself vulnerable to what other people think takes courage.

+ + +

Courage brings us back around to the first foundation, People, because it is the people of the community who will build resilience—and they are the ones who need courage for all the pieces of resilience building:

> Courage to work with other **people** and share in taking responsibility for the community.
>
> Courage to tackle the complex, **systemic** issues we face.
>
> Courage to learn from experience and **adapt** our thinking and methods.
>
> Courage to accept uncertainty and make big **transformations** when necessary.
>
> Courage to commit to far-reaching and long-term resilience building that is truly **sustainable**, for generations to come.

Conclusion

As the United States industrialized in the 19th and 20th centuries, communities had to adapt to the new, modern world. They redesigned streets to accommodate motor vehicles; they regulated sewage and factory waste; they set policies for everything from housing to schools to policing. These adaptations started as local initiatives but quickly spread nationwide as communities copied and adapted them to local needs.[72] In general, they focused on immediate and local problems.

Toward the end of the century a new global environmental awareness emerged, and with it the idea that local actions had global impacts—and vice versa. Communities of all sizes started talking about issues like recycling, economic globalization, and greenhouse gas mitigation. This was a new breed of adaptations to a new set of circumstances, and it was conspicuously marked by the spread of the term "sustainability."

The first fifteen years of the 21st century have underlined the true depth of our communities' sustainability challenges—environmental, economic, and social. The Y2K computer bug scare alerted millions to how vulnerable we now are to disruptions in national and global distribution networks. Hurricane Katrina in 2005 demonstrated how climate change threatens our cities today, not decades in the future. And the oil price spike of 2008 reminded us how dependent we remain on cheap oil.

When the Occupy movement shut down Wall Street in 2011, and then Superstorm Sandy shut down Wall Street a year later, it became clear that the 21st century poses complex challenges unlike those of the last century, and that reach from the smallest town to the heart of global capitalism. It's no surprise that communities have turned to resilience as the best response; resilience is well suited for grappling with the complexity, uncertainty, and multiple scales of these new challenges. Indeed, the spread of the term "resilience" marks the next stage in how we are adapting our communities to new circumstances.

Resilience is, in a way, the original aspiration of human communities. Since the dawn of civilization we have banded together for long-term mutual well-being and betterment in the face of future stresses and shocks. History is full of communities—even highly complex ones—that persisted for thousands of years: they found ways to be resilient despite natural disaster and internal discord, embedding their wisdom and practices in place-based cultures. Of course, history is also full of communities and civilizations that succumbed to external or internal crises, often far larger than they had any possibility of anticipating. While we should heed the warnings of that history, we can also consider ourselves fortunate in the modern era to have a broader view of what crises we might face, and access to countless examples of community resilience both ancient and contemporary. *Six Foundations for Building Community Resilience* aims to help us better understand what made those examples successful, and help existing and future resilience-building efforts across the country be more effective.

Appendix

Six Foundations for Building Community Resilience is the result of a multi-year effort at Post Carbon Institute to explore global-level sustainability challenges and community-level responses through the lens of resilience. It draws upon our past work; the academic literature on resilience; the wisdom of activists, advocates, and professionals working with communities across the country; and the lived experiences of people in communities responding to complex and interrelated challenges.

Reviewers

Hillary Brown, City College of New York

Laird Christensen, Green Mountain College

B. Nick Colby, Swallow Valley Farm

Chuck Collins, Institute for Policy Studies

Michelle Colussi, Canadian Center for Community Renewal

Brooke Deterline, Courageous Leadership

Elsadig Elsheikh, Haas Institute for a Fair and Inclusive Society

Bruce Goldstein, University of Colorado—Boulder

Warren Karlenzig, Common Current

Ann Kinzig, Arizona State University

Michael Lewis, Canadian Center for Community Renewal

Laurie Mazur, Island Press

Leslie Meehan, Thriving Resilient Communities Collaboratory

Max Milton, Alternative Energy Resources Organization

Saharnaz Mirzazad, ICLEI USA

Marissa Mommaerts, Transition U.S.

Judith Gonzalez-Plascencia

Mark Randazzo, Edge Funders

William Rees, co-creator of ecological footprint concept

Jack Santa Barbara, Atamai Village

Philippe Vandenbroeck, shiftN

Andrew Zolli, author of *Resilience: Why Things Bounce Back*

Interviews conducted with advocates and activists

See **resilience.org/talking-resilience** for transcripts of these interviews conducted for this project by Associate Director Ken White

Lynn Benander, Co-op Power

Sarah Byrnes, Institute for Policy Studies

Ellen Choy, Movement Generation

Chuck Collins, Institute for Policy Studies

Timothy DenHerder-Thomas, Cooperative Energy Futures

Trathen Heckman, Daily Acts

Marissa Mommaerts, Transition U.S.

Helena Norberg-Hodge, Local Futures

Vicki Robin, co-author, *Your Money or Your Life*

Doria Robinson, Urban Tilth

Nikki Silvestri, Silvestri Strategies

Major works consulted

Reinette Biggs, Maja Schlüter, and Michael L. Schoon, eds., *Principles for Building Resilience: Sustaining Ecosystem Services in Social-Ecological Systems*, (Cambridge University Press, 2015).

Tom Butler, George Wuerthner, and Daniel Lerch, eds., *The ENERGY Reader: Overdevelopment and the Delusion of Endless Growth*, (Healdsburg, CA: Watershed Media, 2012).

Gillian Bristow and Adrian Healy, "Building Resilient Regions: Complex Adaptive Systems and the Role of Policy Intervention," *Raumforschung und Raumordnung* (2014) 93-102.

Jeremy Caradonna, *Sustainability: A History*, (Oxford University Press, 2014).

Carl Folke, et. al, "Resilience thinking: integrating resilience, adaptability and transformability," *Ecology and Society* 15(4): 20, 2010.

John Foster, *After Sustainability: Denial, Hope, Retrieval*, (New York: Routledge, 2015).

Bruce Goldstein et. al, "Narrating Resilience: Transforming Urban Systems through Collaborative Storytelling," *Urban Studies* October 8, 2013.

Lance Gunderson and C.S. Holling, eds., *Panarchy: Understanding Transformations in Human and Natural Systems*, (Washington, DC: Island Press, 2002).

Richard Heinberg, *Searching for a Miracle: "Net Energy" Limits and the Fate of Industrial Society*, (San Francisco, CA: International Forum on Globalization, 2009).

Richard Heinberg and Daniel Lerch, eds., *The Post Carbon Reader: Managing the 21st Century's Sustainability Crises*, (Healdsburg, CA: Watershed Media, 2010).

Kresge Foundation, *Bounce Forward: Urban Resilience in the Era of Climate Change*, strategy paper with Island Press, 2015.

Daniel Lerch, *Post Carbon Cities: Planning for Energy and Climate Uncertainty*, (Sebastopol, CA: Post Carbon Press, 2007).

Michael Lewis and Pat Conaty, *The Resilience Imperative: Cooperative Transitions to a Steady-State Economy*, (Gabriola Island, BC: New Society Publishers, 2012), p. 25.

Donella Meadows, *Thinking in Systems: A Primer*, (White River Jct., VT: Chelsea Green, 2008).

Marcus Moench, Stephen Tyler, and Jessica Lage, *Catalyzing Urban Climate Resilience: Applying Resilience Concepts to Planning Practice in the ACCCRN Program (2009-2011)*, (ISET- Boulder: Boulder, CO).

Marissa Mommaerts, Ken White, and Ben Roberts, *Weaving the Community Resilience and New Economy Movement: Voices and Reflections from the Field*, (Santa Rosa, CA: Post Carbon Institute, 2014); http://www.resilience.org/resource-detail/2353188-weaving-the-community-resilience-and-new.

Charles Redman, "Should sustainability and resilience be combined or remain distinct pursuits?" *Ecology and Society* 19(2): 37.

Rockefeller Foundation, *City Resilience Framework*, (New York, 2014).

Johan Rockström, et al., "A Safe Operating Space for Humanity," *Nature* 461:24, Sept. 2009.

Jim Thayer, Morgan Rider, and Daniel Lerch, *Resilient against what?: How leading U.S. municipalities are understanding and acting on resilience*, (Santa Rosa, CA: Post Carbon Institute, 2013); http://www.postcarbon.org/publications/resilient-against-what.

Brian Walker and David Salt, *Resilience Practice: Building Capacity to Absorb Disturbance and Maintain Function*, (Washington DC: Island Press, 2012).

Brian Walker and David Salt, *Resilience Thinking: Sustaining Ecosystems and People in a Changing World*, (Washington DC: Island Press, 2012).

Andrew Zolli and Marie Healey, *Resilience: Why Things Bounce Back*, (New York: Free Press, 2012).

Image credits

Page 1: Hoboken ambulance (cc-by-sa) accarrino / flickr; The Ruins of Detroit (cc-by-nc-nd) Trey Campbell / flickr

Page 2: Glacier calving (c) ziggysofi / Shutterstock; Kern County pumpjacks (c) Christopher Halloran; Foreclosure 2008 (cc-by) Jeff Turner / flickr; Black Lives Matter rally (c) Rena Schild / Shutterstock

Page 4: Beartown Woods Natural Area (cc-by) Nicholas A. Tonelli / flickr; Portland skyline (cc-by-nc-sa) C.M. Keiner / flickr

Page 7: Game 5 at Civic Center Plaza (cc-by-nc) eviloars / flickr; block party (cc-by-nc-sa) Eric Allix Rogers / flickr

Page 10: (Montage of images credited for pages 11-26)

Page 11: public-testimony (cc-by) citizenhelder / flickr

Page 14: London, England (c) European Space Agency

Page 17: 2013-04-24-CLUF-Workday, courtesy Mike Bryson, Roosevelt University

Page 20: SoSie+SoSchiff (cc-by-sa) Andrewglaser / flickr

Page 23: Ireland and Wales (cc-by-nc) NASA Marshall Space Flight Center

Page 26: group aerial (cc-by-nc-sa) povertyinitiative / flickr

Page 30: Wharfedale night panorama (cc-by-nc) Tom Blackwell / flickr

Endnotes

---

[1] The dominant focus on climate change in community sustainability and possible directions in which to broaden it are well summarized in the Introduction to Nathan James Bennett, et al., "Communities and change in the anthropocene: understanding social-ecological vulnerability and planning adaptations to multiple interacting exposures," *Reg Environ Change*, published online August 4, 2015; http://link.springer.com/article/10.1007/s10113-015-0839-5.

[2] See Appendix for full list of works consulted.

[3] Johan Rockström, et al., "A Safe Operating Space for Humanity," *Nature* 461:24, September 2009; see also http://www.nature.com/news/specials/planetaryboundaries/index.html. Mathis Wackernagel and William Rees, *Our Ecological Footprint: Reducing Human Impact on the Earth*, (Gabriola Island, BC: New Society Publishers, 1996).

[4] Richard Heinberg, *Searching for a Miracle: "Net Energy" Limits and the Fate of Industrial Society*, (San Francisco, CA: International Forum on Globalization, 2009).

[5] David Fridley, "Nine Challenges of Alternative Energy," in Richard Heinberg and Daniel Lerch, eds., *The Post Carbon Reader: Managing the 21st Century's Sustainability Crises*, (Healdsburg, CA: Watershed Media, 2010).

[6] Richard Heinberg, "Our Renewable Future, Or, What I've Learned in 12 Years Writing about Energy," Post Carbon Institute, January 21, 2015; http://www.postcarbon.org/our-renewable-future-essay/.

[7] Richard Heinberg, *The End of Growth: Adapting to Our New Economic Reality*, (Gabriola Island, BC: New Society Publishers, 2011).

[8] Noah Gordon, "Why Can't People Feel the Economic Recovery?" *The Atlantic*, October 14 2014.

[9] See also Richard Heinberg and Daniel Lerch, eds., *The Post Carbon Reader: Managing the 21st Century's Sustainability Crises* (Healdsburg, CA: Watershed Media, 2010); and William Rees "Cities as Dissipative Structures: Global Change and the Vulnerability of Urban Civilization", in Weinstein and Turner (eds.), *Sustainability science: the emerging paradigm and the urban environment (*New York: Springer, 2012).

[10] Brian Walker and David Salt, *Resilience Thinking: Sustaining Ecosystems and People in a Changing World*, (Washington, DC: Island Press, 2006), p. 1. The socio-ecological system approach to resilience is prominently explored by the international research community represented at Resilience Alliance (http://resalliance.org) and institutions like the Stockholm Resilience Centre (http://stockholmresilience.org).

[11] See Brain Walker and David Salt, *Resilience Practice: Building Capacity to Absorb Disturbance and Maintain Function*, (Washington, DC: Island Press, 2012), Chapter 3. Other authors interpret characteristics or principles of resilience differently, and for different purposes; see, for example, Reinette Biggs, Maja Schlüter, and Michael L. Schoon, eds., *Principles for Building Resilience: Sustaining Ecosystem Services in Social-Ecological Systems* (Cambridge University Press, 2015).

[12] Per Walker and Salt, *Resilience Practice*, p. 5: "In a resilience framework, the concepts of *complex* and *complex systems* carry particular meanings. The three requirements for a complex adaptive system are: it has components that are independent and interacting; there is some selection process at work on those components and on the results of their interactions; variation and novelty are constantly being added to the system (through components changing over time or new ones coming in)."

[13] See for example: My Sellberg, Cathy Wilkinson, and Garry Peterson, "Resilience assessment: a useful approach to navigate urban sustainability challenges" *Ecology and Society* (2015) 20(1):43, http://dx.doi.org/10.5751/ES-07258-200143; Noah Enelow, "The Resilience of Detroit: An Application of the Adaptive Cycle Metaphor to an American Metropolis," *Economics for Equity & Environment*, August 1, 2013, http://www.academia.edu/7973544/The_Resilience_of_Detroit; Rolf Pendall, Kathryn Foster, and Margaret Cowell, "Resilience and regions: building understanding of the metaphor," *Cambridge J Regions Econ Soc* (2010) 3 (1): 71-84, http://www.academia.edu/10183451/Resilience_and_regions_building_understanding_of_the_metaphor; and Nathan James Bennett, et al., "Communities and change in the anthropocene: understanding social-ecological vulnerability and planning adaptations to multiple interacting exposures," *Reg Environ Change*, published online August 4, 2015 at http://link.springer.com/article/10.1007/s10113-015-0839-5.

[14] As philosopher John Foster notes, "mainstreamed as sustainability or sustainable development, environmentalism has failed to reduce, even remotely adequately, the impact of humans on the biosphere"; *After Sustainability: Denial, Hope, Retrieval* (New York: Routledge, 2015), p. 2. For an excellent exploration of sustainability thinking see Jeremy Caradonna, *Sustainability: A History* (Oxford University Press, 2014).

[15] In his review of a draft of this report, William Rees commented: "Resilience planning, emerging from chaos and catastrophe theory, recognizes that the changes coming may be unprecedented and inherently unpredictable... The global systems in which humans are interfering are vastly too complicated for the human mind to understand all possible outcomes, so we must be able to (in [resilience scientist] Buzz Holling's famous words) 'manage for surprise'."

[16] Charles Redman, "Should sustainability and resilience be combined or remain distinct pursuits?," *Ecology and Society* 19(2): 37.

[17] U.S. Supreme Court Justice Louis Brandies noted how a state may "serve as a laboratory; and try novel social and economic experiments without risk to the rest of the country." U.S. Supreme Court, *New State Ice Co. v. Liebmann*, 285 U.S. 262 (1932).

[18] As theoretical physicist Geoffrey West has noted: "One of the great things about being in a city is that there are a lot of crazy people around. I suppose that's another way of saying cities have lots of cognitive diversity...They provide a landscape that allows the spectrum of ideas to blossom. As the city grows, this makes it more and more multidimensional. Cities seem to open up: the spectrum of functionalities, job opportunities, connections, etc. That is key to the vitality and the buzz of successful cities." In Andrew Zolli and Marie Healey, *Resilience: Why Things Bounce Back*, (New York: Free Press, 2012), p. 99.

[19] Among many others, see Paul Willis, "Engaging communities: Ostrom's economic commons, social capital and public relations," *Public Relations Review* 38:1 (March 2012), p.116-122.

[20] Doria Robinson and Ken White, "Living Within a Limit is OK: Talking Resilience with Nikki Silvestri," interview with Post Carbon Institute, June 24, 2015; http://www.resilience.org/stories/2015-06-24/talking-resilience-with-doria-robinson.

[21] From Comstock-Gay's Foreword in: Michael Shuman and Gwendolyn Hallsmith, *Vermont Dollars, Vermont Sense: A Handbook for Investors, Businesses, Finance Professionals, and Everybody Else* (Santa Rosa, CA: Post Carbon Institute, 2015).

[22] Ellen Choy and Ken White, "Talking Resilience with Ellen Choy," interview with Post Carbon Institute, June 24, 2015; http://www.resilience.org/stories/2015-06-24/talking-resilience-with-ellen-choy.

[23] Phil Myrick, "The Power of Place: A New Dimension for Sustainable Development," Project for Public Spaces blog, April 21, 2011; http://www.pps.org/blog/the-power-of-place-a-new-dimension-for-sustainable-development/.

[24] As Bruce Goldstein et. al note, "[I]t is crucial to recognize that urban scales are socially constructed, culturally maintained and politically contested. Cities are relational accomplishments, which matters profoundly to the theorization of resilience for urban city regions." Bruce Goldstein et. al, "Narrating Resilience: Transforming Urban Systems through Collaborative Storytelling" *Urban Studies*, October 8, 2013.

[25] Andrew Zolli notes that "resilience is predicated on trust in a system, allowing potential adversaries to move seamlessly into cooperative mode." *Resilience*, p. 145.

[26] "[W]hat is considered as effective and legitimate adaptation depends on what people perceive to be worth preserving and achieving. How to adapt to climate change therefore hinges on the values underlying people's perspectives on what the goals of adaptation should be." In: Karen O'Brien and Johanna Wolf, "A values-based approach to vulnerability and adaptation to climate change," March 2010, *Wiley Interdisciplinary Reviews: Climate Change* 1(2):232- 242.

[27] See, for example, Thriving Earth Exchange, a project of the American Geophysical Union that makes scientists available to advise communities on climate and natural resource issues: http://thrivingearthexchange.org.

[28] Identity is a complex concept with a long history in the social sciences, including cybernetics; see for example Luc Hoebeke, "Identity: the paradoxical nature of organizational closure", *Kybernetes* (2006) 35(1/2):65-75.

[29] Bruce Goldstein, email to the author, August 19, 2015.

[30] See for example the *Vision PDX* undertaking by the City of Portland, Oregon in 2005-2007. The local government led this multi-year project to develop a community vision for the next twenty years through interviews, surveys, and outreach. http://www.visionpdx.com. See also Goldstein et. al, "Narrating Resilience: Transforming Urban Systems through Collaborative Storytelling" *Urban Studies,* October 8, 2013.

[31] Of the many conceptualizations of social capital, a commonly used one is from Robert Putnam, author of the classic *Bowling Alone*: "[S]ocial capital refers to features of social organization such as networks, norms and trust that facilitate co-ordination and co-operation for mutual benefit." Robert Putnam, *Making Democracy Work: Civic Traditions in Modern Italy*, (Princeton, NJ: Princeton University Press, 1993), p. 35.

[32] As Michael Lewis and Pat Conaty note, " ...centralized, distant, and locally unaccountable power cannot accomplish the transition to low-carbon, ecologically sustainable communities. ...[R]esilience requires a quality of social capital – trust, collaboration, cooperation, and leadership– rooted in the places where people live.": Michael Lewis and Pat Conaty, *The Resilience Imperative: Cooperative Transitions to a Steady-State Economy*, (Gabriola Island, BC: New Society Publishers, 2012), p. 25. See also Jeffrey Potent, "Employing a Knowledge Systems Approach to Creating a Sustainable Future," State of the Planet blog, Earth Institute, Columbia University, February 21, 2014; http://blogs.ei.columbia.edu/2014/02/21/employing-a-knowledge-systems-approach-to-creating-a-sustainable-future.

[33] "[T]he networked nature of governance and policy is critical in resilience," governance being understood as emerging from the interactions of many public and private actors of which government is but one. Gillian Bristow and Adrian Healy, "Building Resilient Regions: Complex Adaptive Systems and the Role of Policy Intervention," *Raumforschung und Raumordnung* (2014) 93-102.

[34] Trevor Tompson, et al., *Resilience in the Wake of Superstorm Sandy*, (The Associated Press-NORC Center for Public Affairs Research, June 2013); http://www.apnorc.org/projects/Pages/resilience-in-the-wake-of-superstorm-sandy.aspx.

[35] As Wes Jackson notes, "I think there's a general law: High energy destroys information, of a cultural as well as a biological variety. There's a loss of cultural capacity. And from 1750, the beginning of the Industrial Revolution, the graphical curve for the use of high-energy fossil carbon is increasingly steep." Wes Jackson, "Tackling the Oldest Environmental Problem: Agriculture and Its Impact on Soil," in Richard Heinberg and Daniel Lerch, eds., *The Post Carbon Reader: Managing the 21st Century's Sustainability Crises*, (Healdsburg, CA: Watershed Media, 2010); http://www.postcarbon.org/publications/agriculture-and-its-impact-on-soil.

[36] Michelle Colussi, email to author, August 19, 2015.

[37] Nikki Silvestri and Ken White, "Economy, Ecology, Equity: Talking Resilience with Nikki Silvestri," interview with Post Carbon Institute, March 31, 2015; http://www.resilience.org/stories/2015-03-31/talking-resilience-nikki-silvestri.

[38] Many primers on systems thinking exist; a fairly accessible one is Donella Meadows, *Thinking in Systems: A Primer*, (White River Jct., VT: Chelsea Green, 2008).

[39] Walker and Salt, *Resilience Practice*, p. 5.

[40] "There are known knows," Wikipedia, accessed August 19, 2014; https://en.wikipedia.org/wiki/There_are_known_knowns.

[41] William Rees, "Cities as Dissipative Structures." J Moore and W. Rees, "Getting to One Planet Living," in *State of the World 2013 – Is Sustainability still Possible?* (Washington DC:, Island Press, 2013).

[42] David Fridley, "Nine Challenges of Alternative Energy," in Richard Heinberg and Daniel Lerch, eds., *The Post Carbon Reader: Managing the 21st Century's Sustainability Crises*, (Healdsburg, CA: Watershed Media, 2010).

[43] See C.S. Holling, Lance Gunderson, and Garry Peterson, "Sustainability and Panarchies" in Gunderson and Holling, eds., *Panarchy: Understanding Transformations in Human and Natural Systems*, (Washington, DC: Island Press, 2002).

[44] Eric Hoffer, *Reflections on the Human Condition* (Harper & Row, 1973).

[45] Walker and Salt, *Resilience Practice*, p 92-98.

[46] Reinette Biggs; Maja Schlüter; Michael Schoon, eds., *Principles for building resilience : sustaining ecosystem services in social-ecological systems*, (Cambridge, UK: Cambridge University Press, 2015). Stockholm Resilience Institute has a useful summary at http://www.stockholmresilience.org/21/research/research-news/2-19-2015-applying-resilience-thinking.html.

[47] Rockefeller Foundation, *City Resilience Framework*, (New York, 2014); https://www.rockefellerfoundation.org/report/city-resilience-framework.

[48] Andrew Zolli, "The Verbs of Resilience," personal blog, October 28, 2013; http://andrewzolli.com/the-verbs-of-resilience.

[49] Chicago's infamous Cabrini-Green Homes are a textbook example of the social and economic problems that have at times been exacerbated by "efficient" public housing; https://en.wikipedia.org/wiki/Cabrini%E2%80%93Green_Homes.

[50] See, for example, James Hamilton, "Oil prices and the economic recession of 2007-08," Centre for Economic Policy Research blog, June 16, 2009; http://www.voxeu.org/article/did-rising-oil-prices-trigger-current-recession. For a much broader perspective, including the example of the "2008 financial-energy crisis," see Thomas Homer-Dixon, et al., "Synchronous failure: the emerging causal architecture of global crisis," *Ecology and Society* 20(3):6 (2015).

[51] Of course, this does not imply that this system has been fair or beneficial, or that it will be resilient over the long-term.

[52] Brian Walker, "The best explanation to resilience," Stockholm Resilience Centre TV, uploaded April 3, 2009; https://www.youtube.com/watch?v=tXLMeL5nVQk (minute 5:33).

[53] Giuseppe di Lampedusa, *The Leopard* (New York: Pantheon, 2007 edition), translated by Archibald Colquhuon.

[54] Walker and Salt, *Resilience Practice*, Chapter 3.

[55] Laurie Guevara-Stone, "A High-Renewables Tomorrow, Today: Güssing, Austria," Rocky Mountain Institute blog, October 8, 2013; http://blog.rmi.org/blog_2013_10_08_high-renewables_tomorrow_today_gussing_austria.

[56] These attributes of transformability are drawn from Walker and Salt, *Resilience Practice*, p. 100-103.

[57] Walker & Salt, *Resilience Practice*, p. 101.

[58] See Paul Schmitz, "How Change Happens: The Real Story of Mrs. Rosa Parks and the Montgomery Bus Boycott," *Huffington Post*, December 1, 2014.

[59] Jeremy Caradonna, *Sustainability: A History* (Oxford University Press, 2014), p. 5.

[60] Richard Heinberg, "Beyond the Limits to Growth," in Richard Heinberg and Daniel Lerch, eds., *The Post Carbon Reader: Managing the 21st Century's Sustainability Crises* (Healdsburg, CA: Watershed Media, 2010).

[61] Global Footprint Network, http://www.footprintnetwork.org.

[62] This is an example of "strong sustainability" versus "weak sustainability" as understood in the field of ecological economics.

[63] Johan Rockström, et al., "A Safe Operating Space for Humanity," *Nature* 461:24, September 2009; see also http://www.nature.com/news/specials/planetaryboundaries/index.html.

[64] Richard Heinberg, "What is Sustainability?," Richard Heinberg and Daniel Lerch, eds., *The Post Carbon Reader: Managing the 21st Century's Sustainability Crises*, (Healdsburg, CA: Watershed Media, 2010).

[65] Ibid.

[66] For example, John Foster lambastes "the distracting late-twentieth-century mindset of 'sustainable development', with its obsessive focus on inherently negotiable futures"; *After Sustainability: Denial, Hope, Retrieval* (New York: Routledge, 2015), p. 12.

[67] This sentiment was epitomized in the iconic little book, *50 Simple Things You Can Do to Save The Earth* (Earthworks Group, 1989).

[68] Michelle Colussi, email to author, August 19, 2015.

[69] Vaclav Havel, *Disturbing the Peace: A Conversation with Karel Hvizdala*, English translation by Paul Wilson (New York: Vintage, 1991).

[70] Email with the author, October 28, 2015.

[71] Jenny Leis and Daniel Lerch, eds., *City Repair's Placemaking Guidebook*, (City Repair and Southeast Uplift: Portland, OR, 2003), p. 61.

[72] For example, the use of zoning codes to separate conflicting land uses (like factories and residences) emerged from a handful of communities in the early 1900s and went on to dominate American city planning for most of the century; but by the end of the century, a "New Urbanist" movement had emerged to adapt planning practice in light of the problems unwittingly created by single-use zoning.

Made in the USA
Monee, IL
05 November 2026

67377159R00029